Disclaimer

The publisher of this book is by no way associated with the National Institute of Standards and Technology (NIST). The NIST did not publish this book. It was published by 50 page publications under the public domain license.

50 Page Publications.

Book Title: WWVB Radio Controlled Clocks: Recommended Practices for Manufacturers and Consumers (2009 edition)

Book Author: Michael A. Lombardi; Andrew N. Novick; John P. Lowe; Matthew J. Deutch; Glenn K. Nelson; Douglas D. Sutton; William C. Yates; D. W. Hanson

Book Abstract: Radio controlled clocks represent a true revolution in timekeeping. Clocks that synchronize to NIST radio station WWVB now number in the millions in the United States, and new sales records are being established every year. As a result, many of us are now accustomed to having clocks in our homes, offices, and on our wrists that always display the correct time and that never require adjustment. This NIST Recommended Practice Guide was written to provide guidance to both manufacturers and consumers of radio controlled clocks. Through voluntary compliance with the recommended practices listed here, manufacturers can benefit by continuing to develop more reliable and usable radio controlled products, increasing both consumer confidence and sales. Consumers can benefit by using this guide to help them select and purchase radio controlled clock products, to learn how the products work, and to help troubleshoot reception problems.

Citation: NIST SP - 960-14-09

Keyword: radio controlled clocks;time;time-of-day;WWVB

practice guide

WWVB Radio Controlled Clocks: Recommended Practices for Manufacturers and Consumers
(2009 Edition)

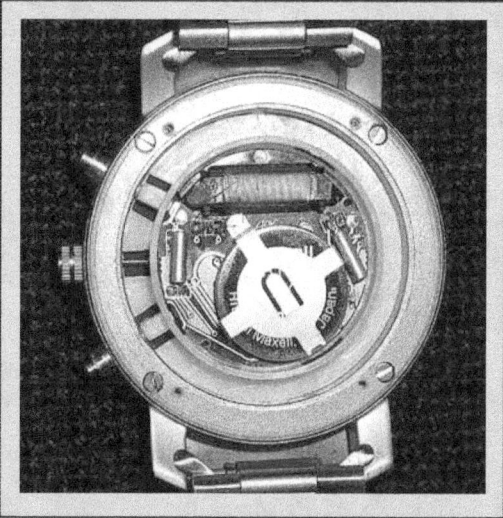

Michael A. Lombardi, Andrew N. Novick,
John P. Lowe, Matthew J. Deutch,
Glenn K. Nelson, Douglas S. Sutton,
William C. Yates, and D. Wayne Hanson

National Institute of
Standards and Technology
U.S. Department of Commerce

Special Publication 960-14

NIST Recommended Practice Guide

Special Publication 960-14

WWVB Radio Controlled Clocks: Recommended Practices for Manufacturers and Consumers (2009 Edition)

Michael A. Lombardi, Andrew N. Novick,
John P. Lowe, Matthew J. Deutch,
Glenn K. Nelson, Douglas S. Sutton,
and William C. Yates
NIST Physics Laboratory

D. Wayne Hanson
Time Signal Engineering

August 2009

U.S. Department of Commerce
Gary Locke, Secretary

National Institute of Standards and Technology
Patrick D. Gallagher, Deputy Director

Certain commercial entities, equipment, or materials may be identified in
this document in order to describe an experimental procedure or concept
adequately. Such identification is not intended to imply recommendation or
endorsement by the National Institute of Standards and Technology, nor is it
intended to imply that the entities, materials, or equipment are necessarily the
best available for the purpose.

National Institute of Standards and Technology
Special Publication 960-14
Natl. Inst. Stand. Technol.
Spec. Publ. 960-14
64 pages (August 2009)
CODEN: NSPUE2

U.S. GOVERNMENT PRINTING OFFICE
WASHINGTON: 2009

For sale by the Superintendent of Documents
U.S. Government Printing Office
Internet: bookstore.gpo.gov   Phone: (202) 512–1800   Fax: (202) 512–2250
Mail: Stop SSOP, Washington, DC 20402-0001

FOREWORD

Radio controlled clocks represent a true revolution in timekeeping. Clocks that synchronize to NIST radio station WWVB now number in the millions in the United States, and new sales records are being established every year. As a result, many of us are now accustomed to having clocks in our homes, offices, and on our wrists that always display the correct time and that never require adjustment. This *NIST Recommended Practice Guide* was written to provide guidance to both manufacturers and consumers of radio controlled clocks. Through voluntary compliance with the recommended practices listed here, manufacturers can benefit by continuing to develop more reliable and usable radio controlled products, increasing both consumer confidence and sales. Consumers can benefit by using this guide to help them select and purchase radio controlled clock products, to learn how the products work, and to help troubleshoot reception problems.

WWVB Radio Controlled Clocks

Acknowledgments

This booklet is a revised version of the original booklet published in January 2005. Although not a major revision, it updates information about radio station WWVB and the United States daylight saving time rules that were changed in 2007. It also makes a few small corrections to the time zone tables and maps. In addition, the recommended practices for radio controlled clock manufacturers have been revised slightly to incorporate new information collected by the authors since the original booklet was published.

The authors gratefully acknowledge those who reviewed this document and made numerous helpful suggestions including: Michael Chiu and Hans-Joachim Sailer of C-Max Technology; Etsuro Nakajima and Aihara Fumikazu of the Casio Corporation; Glenn Burdett of the Spectracom Corporation; Rod Mack of Ultralink; John Rowland and Gene Fornario, the founder and co-moderator respectively of the Casio Waveceptor forum on Yahoo.com; Tom O'Brian, Chief of the NIST Time and Frequency Division; and David Smith, Chairman of the Editorial Review Board at NIST's Boulder laboratories. We also thank all those who have called and e-mailed NIST in recent years and shared their personal experiences with radio controlled clock (RCC) products. Their questions and comments inspired us to create this guide.

WWVB Radio Controlled Clocks

Table of Contents

List of Figures x

List of Tables xi

1. Introduction 1

2. Technical Description of WWVB 3

3. Recommended Practices for Clock Accuracy,
   Clock Display, and Controls 5

   3.A. Analog Clock Displays 5

   3.B. Digital Clock Displays 6

   3.C. RCC Controls 7

   3.D. Compatibility with Other Stations 7

   3.E. Signal Quality Indicator 8

   3.F. Antenna Orientation 9

4. Recommended Practices for Clock Synchronization 11

   4.A. Initial Synchronization When Clock
        Is First Turned On or Reset 11

      4.A.1. Analog Clocks (Hand Alignment) 11

      4.A.2. Digital Clocks 11

   4.B. Synchronization by Radio at Assigned Times 11

      4.B.1. Amount of Time Allotted to a
             Synchronization Attempt 14

      4.B.2. Status of Display During a
             Synchronization Attempt 14

4.C. Synchronization by Radio at a
Time Selected by the User 14

4.D. Manual Clock Synchronization 14

4.E. Synchronization Indicator..15

    4.E.1. Digital Clock Synchronization Indicator 16

    4.E.2. Analog Clock Synchronization Indicator 16

4.F. Adjustment of Display to Compensate
for Delays Introduced During Synchronization...............................17

5. Recommended Practices for
Time Zone Settings 19

   5.A. Time Zone Selection 19

6. Recommended Practices for
Daylight Saving Time (DST) 27

   6.A. Handling of Transition Days...27

   6.B. Disabling/Enabling DST Switch 28

   6.C. DST Indicator 28

7. Recommended Practices for Leap Seconds,
Leap Years, and the Two-Digit Year Code 29

   7.A. Handling of Leap Seconds 29

   7.B. Handling of Leap Years 29

   7.C. Handling of Two-Digit Year Code 30

8. Recommended Practices for
Hardware Specifications 31

   8.A. Receiver Specifications..31

8.B.	Antenna Considerations		31
8.C.	Local Oscillator Specifications		32
8.D.	Battery Powered RCCs		32

9. Recommended Practices for Product Documentation 33

    9.A. Mention of NIST 33

    9.B. Use of "Atomic Clock" Nomenclature 34

10. Compliance Checklist 35

11. Recommended Practices for Consumers of WWVB RCCs 41

    11.A. How a WWVB RCC Works 41

    11.B. Time Zone Settings 42

    11.C. Coverage Area of the WWVB Signal 43

    11.D. General Troubleshooting Tips for WWVB RCCs 44

        11.D.1. General Troubleshooting Tips for RCCs That Won't Synchronize at All 45

        11.D.2. General Troubleshooting Tips for RCCs Off by One Hour or More 48

        11.D.3. General Troubleshooting Tips for RCCs Off by a Few Minutes or Seconds 48

        11.D.4. General Troubleshooting Tips Concerning Daylight Saving Time (DST) 49

12. References 51

List of Figures

Figure 1. WWVB time code format 3

Figure 2. Analog clock display with digital inset for date information 5

Figure 3. Digital clock display with date information 6

Figure 4. Digital clock display that includes year information 6

Figure 5. A three-segment display indicating four levels of signal quality 8

Figure 6. Use of a broadcast tower icon as a signal quality meter 8

Figure 7. RCC display indicating the date and time of the
last synchronization 15

Figure 8. World time zone map 25

Figure 9. United States time zone map 26

List of Tables

Table 1. Radio Time Signal Stations Used for RCC Synchronization 7

Table 2. Dark Path Hours and Duration for WWVB Signal at
Selected Cities (Local Time) 12

Table 3. Necessary Time Zone Options for WWVB RCC Products 20

Table 4. Recommended Time Zone Options for WWVB RCC
Products Sold Internationally 21

Table 5. Difference between UTC and Local Time for the Four Major
Time Zones in the CONUS 43

Table 6. Potential Sources of Interference for WWVB RCCs 46

WWVB Radio Controlled Clocks

1. INTRODUCTION

Through its Time and Frequency Division located in Boulder, Colorado, the National Institute of Standards and Technology (NIST) maintains official time and frequency standards for the United States of America and distributes these standards to the American public. Radio station WWVB, located near Fort Collins, Colorado, is one of the most important distribution sources for these standards. The station continuously broadcasts a 70 kW signal at a frequency of 60 kHz that covers the Continental United States (CONUS), and also reaches Alaska and Hawaii during the nighttime hours.

The signals from WWVB can serve as a convenient reference standard for time interval and frequency, but their primary function is the time-of-day synchronization of radio controlled clocks (RCCs). These clocks are now sold through a variety of retail channels to United States consumers, who install them in their homes and offices and rely on them as reliable and official sources of time, accurate to within 1 second (s) or less. These clocks are sold in a variety of forms, as wall clocks, desk clocks, wristwatches, or clocks embedded into a variety of consumer electronic products, including kitchen appliances such as coffee makers and microwave ovens, home entertainment equipment, and computer systems.

As with all consumer electronic products, the quality of WWVB RCCs produced by different manufacturers varies widely. While many of the existing products are well designed and extremely reliable, some models cannot always decode the time signal even under the most favorable signal conditions, and some lack key features that limit their usefulness from a human engineering standpoint. NIST has produced this recommended practice guide for the benefit of manufacturers and consumers of WWVB radio controlled clocks. It recommends key features to manufacturers that their products should include, as well as key specifications that their products should meet.

These recommended practices are voluntary. No manufacturer or consumer is required by law to comply with them. However, it is hoped that voluntary compliance with these recommended practices will lead to the development of more reliable and usable clocks, increase consumer confidence in WWVB RCCs, and, at the same time, increase the size of the commercial RCC marketplace in the United States. It is also hoped that this guide will be useful to consumers of WWVB RCCs by providing information that helps them select RCC products and troubleshoot RCC reception problems.

The following sections contain a brief technical description of WWVB (Section 2) and product recommendations in seven different categories, including: clock displays (Section 3), synchronization (Section 4), time zone settings (Section 5), handling of daylight saving time (Section 6), handling of miscellaneous time code issues (Section 7), hardware specifications (Section 8), and product documentation (Section 9). Based on consumer feedback received at NIST, the seven categories were identified as being important to the reliability, usability, and marketability of WWVB RCC products. Not all categories share equal importance, so a summary checklist is provided in Section 10 that identifies which categories are recommended as necessary for all designs and which categories are optional. However, we recommend that manufacturers comply with as many of these optional categories as possible in an effort to produce RCC products of the highest quality.

Many consumers will find the information in Sections 2 to 10 interesting, but some will probably want to skip directly ahead to Section 11, which is included for their benefit. It provides general information about WWVB RCCs, and it provides some troubleshooting tips for consumers whose clocks are not working properly.

Please note that it was necessary to write this guide in a general fashion. We identify for manufacturers the desired functions of WWVB RCC products, but a discussion of the technical implementation of these functions is beyond the scope of this guide.

2. TECHNICAL DESCRIPTION OF WWVB

The WWVB time code includes 60 bits of information, transmitted at 1 bit per second. A full minute (60 s) is required to send a complete time code frame (Figure 1). An on-time marker (OTM) is sent every second by reducing the power of the 60 kHz carrier frequency by 17 dB at a time that coincides with the arrival of the Coordinated Universal Time (UTC) second. Bits are identified by the length of time that the carrier power is held low. A 0 bit is sent by holding the power low for 200 ms, a 1 bit is sent by holding the power low for 500 ms. Frame markers are sent every 10 s by holding the power low for 800 ms.

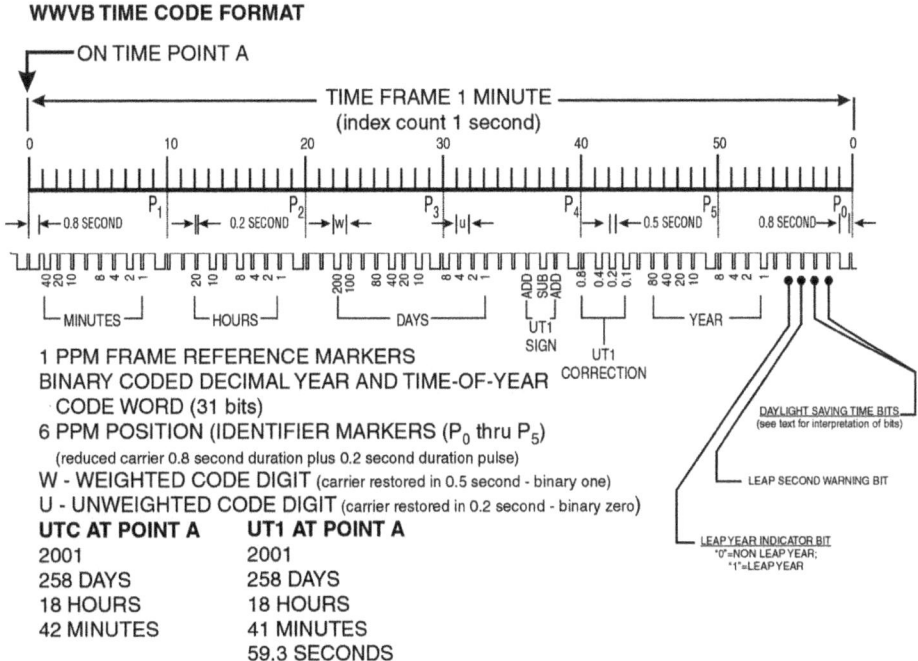

Figure 1. WWVB time code format.

A second time scale is located at the radio station site in Fort Collins, Colorado. The two time scales, both consisting of groups of atomic oscillators, are continuously compared to each other and kept in very close agreement. The time difference between them is nearly always less than 20 nanoseconds.

3

A complete technical description of the WWVB and its broadcast format is not provided here. A general technical description can be found in *NIST Special Publication 432*,[1] and a more detailed description of the station operation, format, and broadcast control is available in *NIST Special Publication 250-67*.[2] Both publications are available for download from the NIST Time and Frequency Division web site at **http://tf.nist.gov**

3. RECOMMENDED PRACTICES FOR CLOCK ACCURACY, CLOCK DISPLAY, AND CONTROLS

All RCC products should display time accurate to at least within ±0.5 s, so that when the time is rounded to the nearest second, the seconds' value is always correct. Tighter synchronization (to within ±0.2 s) is desirable. This prevents the human eye from detecting any errors when checking a RCC display against another independent time reference, whereas a 0.5 s error could be noticeable.

The chief benefit and a key selling point of a RCC is its time accuracy. Therefore, we recommend that all clocks display seconds, or have the option of displaying seconds. With some digital clocks, such as clock radios, the seconds' digits can be made smaller than the hour and minute digits, or the consumer should be given the option to disable the seconds' display if they find it distracting. Analog clocks require either a second hand or a separate digital display that shows seconds. A digital seconds' display that can be turned on and off is often a good option because second hands are not always desirable on analog clocks. For example, a second hand on an analog alarm clock might be noisy enough to bother consumers who are trying to sleep.

3.A. Analog Clock Displays

Analog clock displays should include an hour, minute, and second hand (or a digital display of seconds), and some marking or label indicating that the clock is radio controlled. The date and other information can be displayed (if desired) in a digital inset, as shown in Figure 2.

Figure 2. *Analog clock display with digital inset for date information.*

3.B. Digital Clock Displays

Digital clock displays should display the hour, minute, second, and some marking or label indicating that the clock is radio controlled. The digital display makes it convenient to also display the month, day, year, and weekday (if desired). It is also recommended that "AM" or "PM" is displayed if the clock is set to a 12-hour format, instead of a 24-hour format. Some digital clock displays use an icon of a satellite dish as a synchronization indicator (Section 4.E), as a signal quality indicator (Section 3.E), or simply to indicate that the clock is radio controlled. To avoid confusing consumers, we recommend that a picture of a satellite dish not be used, since WWVB and the other low-frequency (LF) time-signal stations (Section 3.D) all broadcast from ground-based transmitters and not from satellites. Sample digital clock displays are shown in Figures 3 and 4.

Figure 3. Digital clock display with date information.

Figure 4. Digital clock display that includes year information.

3.C. RCC Controls

RCC controls should be clearly labeled and situated in such a way that they cannot be accidentally activated; for example, everyday handling of the clock should not result in the change of a time zone setting. This is particularly important in the case of wristwatches.

3.D. Compatibility with Other Stations

Some WWVB RCC products are capable of receiving time signals broadcast by other time signal stations located in other countries. This allows the clocks to be sold and used internationally. The time signals broadcast by other countries (Table 1) use carrier frequencies different from that of WWVB in some cases, and different time code formats in all cases, but the modulation schemes are similar. We recommend that products capable of receiving more than one time signal have a way of clearly indicating to the consumer which time signal is currently being received. We also recommend that RCCs are designed to change stations automatically (if necessary) when the time zone setting is changed (Section 5.A). For example, if a person wearing a wristwatch with multi-station capability travels from the United States

Table 1: **Radio Time Signal Stations Used for RCC Synchronization**

Station Call Sign	Country	Controlling Agency	Carrier Frequency
WWVB	United States	National Institute of Standards and Technology (NIST)	60 kHz
BPC	China	National Time Service Center (NTSC), Chinese Academy of Sciences	68.5 kHz
DCF77	Germany	Physikalisch-Technische Bundesanstalt (PTB)	77.5 kHz
HBG	Switzerland	Swiss Federal Office of Metrology and Accreditation (METAS)	75 kHz
JJY	Japan	National Institute of Information and Communications Technology (NICT)	40 kHz, 60 kHz
MSF	United Kingdom	National Physical Laboratory (NPL)	60 kHz

to Germany, the wristwatch should automatically switch from WWVB to DCF77 when the German time zone is selected. In areas covered by more than one station, such as Japan, the RCC should automatically select the station with the highest signal quality.

3.E. Signal Quality Indicator

Inclusion of a real-time signal quality indicator is recommended so that the consumer can find the best location and antenna orientation for their RCC product while forcing the product to attempt synchronization (Section 4.C). Since the actual signal strength is not easy to detect due to other RF noise at 60 kHz, the signal quality indicator can show the "bit strength," or current readability level of the signal; or it can indicate the progress of the decoding process in the software.

The signal quality indicator can be simple. A three-segment display indicating a low-, medium-, and high-quality signal is generally adequate (Figures 5 and 6). When the clock is not attempting to synchronize and

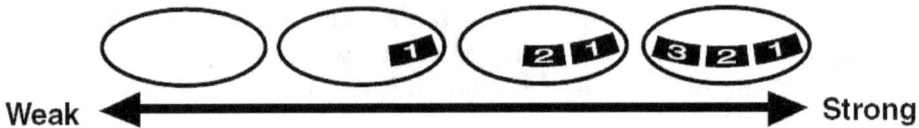

Weak ⬅————————————————➡ Strong

Figure 5. A three-segment display indicating four levels of signal quality.

 📶 (strong)	Strong
 📶 (weak)	Weak
 📶 (none)	No Reception
 📶 (flashing)	Receiving

Figure 6. Use of a broadcast tower icon as a signal quality meter.

the radio receiver is turned off or disabled, we recommend that the signal quality indicator also be disabled or removed from the display. Otherwise, consumers might mistakenly assume that the RCC is displaying the current signal quality.

3.F. Antenna Orientation

Most RCC antennas are directional and achieve maximum gain when they are positioned broadside to the transmit antenna in Fort Collins, Colorado. We recommend that an arrow or pointer marker is included on the RCC case to illustrate the antenna orientation. When used in conjunction with the signal quality indicator (Section 3.E), this type of marking can assist consumers in orienting the RCC to obtain maximum signal strength.

4. RECOMMENDED PRACTICES FOR CLOCK SYNCHRONIZATION

This section recommends practices for clock synchronization. It is divided into four categories, initial synchronization (when the clock is first turned on), synchronization by radio at assigned times, synchronization by radio at a time selected by the consumer, and manual synchronization without radio control.

4.A. Initial Synchronization When Clock Is First Turned On or Reset

When a RCC is first turned on, it will begin looking for a signal and attempt to synchronize. We recommend that RCCs be designed to continuously try to synchronize on this first attempt until either the synchronization is successful or until the consumer decides to attempt manual synchronization (Section 4.D). We also recommend that RCCs that have not yet been able to synchronize should not run or attempt to display the time since their displays will be incorrect. Suggestions for what a RCC should display prior to synchronization are listed below.

All RCCs should be designed to synchronize without any interaction from the user. It should not be necessary for the user to set the display of the clock (move the hands, etc.) in order for the clock to synchronize. We recommend that at least two consecutive time codes are decoded and compared before determining that an initial synchronization attempt is successful.

4.A.1. Analog Clocks (Hand Alignment)

Analog clocks should align themselves with all hands pointed upward (pointed to the "12") until synchronization is successful. The hands should remain motionless during synchronization.

4.A.2. Digital Clocks

Digital clocks should display 12:00:00 as the time prior to synchronization, or flash the time display on and off, or display dashes instead of the hours, minutes, and seconds. The display should not increment during synchronization.

4.B. Synchronization by Radio at Assigned Times

To meet the accuracy requirements listed in Section 3, and to periodically check the time code for notifications of daylight saving time, leap seconds, and other time code changes, all WWVB RCCs should attempt to synchronize at least once every 24 hours, and more frequently if possible.

If only one synchronization attempt is made, it should be made at night when the signal is the strongest. The signal is generally easiest to receive when it is dark at both the transmitter site in Fort Collins, Colorado, and at the site where the RCC is located, so the highest probability of a successful synchronization is during these hours. Table 2 shows the dark path hours (DPH) and dark path duration (DPD) in hours and minutes for six cities on the approximate longest and shortest days of 2004 (June 21 and December 21), based on when sunrise and sunset occurs in those cities with respect to when sunrise and sunset occur in Fort Collins. The DPH are based on the local time in the selected city. The six cities were chosen to represent the northwest, southwest, northeast, and southeast corners of the CONUS, as well as Alaska and Hawaii.

The information in Table 2 shows that the "window of opportunity" for synchronization ranges from about 4 hours (Anchorage summer) to about 14 hours (Seattle winter). Attempting synchronization on the hour at midnight, 1 a.m., and 2 a.m. guarantees a dark path at all United States

*Table 2: **Dark Path Hours and Duration for WWVB Signal at Selected Cities (Local Time)***

	June 21, 2004		December 21, 2004	
City	DPH	DPD	DPH	DPD
Seattle, Washington	9:11 p.m. to 4:30 a.m.	7:19	4:20 p.m. to 6:21 a.m.	14:01
San Diego, California	8:00 p.m. to 4:30 a.m.	8:30	4:46 p.m. to 6:21 a.m.	13:35
Caribou, Maine	10:34 p.m. to 4:38 a.m.	6:04	6:36 p.m. to 7:14 a.m.	12:38
Miami, Florida	10:34 p.m. to 6:30 a.m.	7:56	6:36 p.m. to 7:03 a.m.	12:27
Anchorage, Alaska	11:42 p.m. to 3:30 a.m.	3:48	3:41 p.m. to 5:21 a.m.	13:40
Honolulu, Hawaii	7:16 p.m. to 2:30 a.m.	7:14	5:55 p.m. to 4:21 a.m.	10:26

locations. Therefore, if only one synchronization attempt is made, we recommend that the scheduled synchronization time is set to be midnight, 1 a.m., or 2 a.m. (local receiver time).

Manufacturers are cautioned to carefully choose a radio synchronization time and method that does not cause the RCC to display the wrong time (even briefly) on transition days to and from daylight saving time, when the time display should be adjusted forwards or backwards by one hour at exactly 2 a.m. local time (Section 6.A).

If the receiver and signal processing firmware (Section 8.A) inside the RCC are of sufficient quality, it should be possible for successful synchronizations to be routinely made throughout the CONUS during both the daytime and nighttime hours. Therefore, we recommend that manufacturers design products that attempt synchronization more than once per day. This has the disadvantage of perhaps reducing the battery life on battery-powered devices, but it has the advantage of relaxing the quartz oscillator accuracy requirements (Section 8.C) by shortening the interval between synchronizations. For example, if the RCC is programmed to synchronize at both midnight and 4 a.m., it reduces the amount of time that the clock free runs on its local oscillator from 24 to 20 hours, relaxing the oscillator specification by roughly 16%.

If the quartz oscillator is of sufficient quality to maintain accurate time for 24 hours (Section 8.3), the manufacturer may elect to offer several scheduled synchronization times, but to skip the times remaining on the schedule once a successful synchronization has been made. For example, a RCC could attempt to synchronize at midnight, 2 a.m., 4 a.m., and 6 a.m. If the synchronization at midnight is successful, it can skip the attempts at 2 a.m., 4 a.m., and 6 a.m., and wait until the following midnight before trying again. However, if the midnight attempt fails, the RCC will try again at 2 a.m. instead of waiting for 24 hours.

Since the small antennas used by WWVB RCCs tend to be very directional, it is difficult for a RCC to synchronize while it is moving. Therefore, it probably won't be advantageous for wristwatches to attempt to synchronize during the daytime hours when the watch is being worn and is probably in motion. As a result, we recommend that wristwatches attempt their multiple synchronizations at times when the consumer is most likely to be asleep, and the watch is motionless. Several synchronization attempts during the night are recommended to allow for the varying "bed times" of consumers.

Manufacturers should be aware of potential problems introduced by consumers who work nights or who never remove their watches. To meet the needs of these consumers, manufacturers can elect to allow an additional synchronization time (in addition to those built-in to the product) to be manually selected. If this is done, the product documentation (Section 9) should adequately explain to the consumer when the product is most likely to successfully synchronize.

4.B.1. Amount of Time Allotted to a Synchronization Attempt

We recommend that WWVB RCCs attempt to decode time codes for at least five consecutive minutes before determining that a synchronization attempt has failed.

4.B.2. Status of Display During a Synchronization Attempt

During the synchronization attempt it may be necessary to partially disable or alter the RCC's display. For example, the second hand on an analog clock might stop, or the display on a digital clock might "flash" or show a message indicating that it is now receiving a radio signal. However, it is very important that the RCC continues to keep time during the synchronization attempt. If the attempt fails, the clock can then restore the time display and retain the same accuracy that it had prior to the failed attempt.

4.C. Synchronization by Radio at a Time Selected by the User

All WWVB RCCs should include a button or control that allows the user to attempt immediate synchronization, without waiting for the next scheduled synchronization. The RCC will attempt to decode the incoming signal whenever this control is activated. The RCC display should indicate whether the synchronization attempt succeeded or failed.

This button or control should be designed so that it cannot be activated by accident if the RCC is bumped or jostled. To avoid accidental activation, it might be desirable to use a button that needs to be held in place for several seconds before the clock attempts synchronization.

4.D. Manual Clock Synchronization

All WWVB RCCs should allow the user to disable radio controlled timekeeping functions, so they can be operated as conventional clocks if necessary. This means that the consumer should be allowed to manually synchronize the time and date settings if the signal is unreceivable. This protects the RCC from becoming obsolete if it is moved to an area outside of the signal range. The display should indicate if the RCC is being operated without radio control.

4.E. Synchronization Indicator

All WWVB RCCs should indicate whether they have recently synchronized. Since the clocks are radio controlled and advertised as accurate, they are trusted by consumers, who typically assume that the displayed time is exactly right. However, this will not be true if the clock has not received the signal for a long period. All manufacturers should realize that it is extremely important to communicate to the person viewing the clock that the time can be trusted. This requires indicating whether or not the RCC has been recently synchronized.

Ideally, the RCC should be able to display the date and time of the last synchronization. If this is not possible, the RCC should indicate in some fashion whether it has been more than 24 hours since the last synchronization or, preferably, the total length of time (probably expressed in days) since the last synchronization. Since manufacturers should expect their products to work when used within the coverage area,

Figure 7. RCC display indicating the date and time of the last synchronization.

it is more appropriate for RCCs to indicate when synchronization has failed in the last 24 hours (an abnormal condition), than it is to indicate when synchronization has succeeded (a normal condition). Examples of appropriate synchronization indicators for digital and analog clocks are provided in Sections 4.E.1. and 4.E.2.

Some manufacturers may elect to only alert the consumer if synchronization has not occurred in a period longer than 24 hours; for example, 48 hours or 72 hours could be used. However, all manufacturers should attempt to meet the Section 3 requirement of keeping time between synchronizations to within ±0.5 s of UTC(NIST). A longer interval should be used only if the local oscillator (Section 8.C) is capable of keeping time to within ±0.5 s of UTC(NIST) throughout the entire interval. For this reason, synchronization intervals longer than 24 hours might not be acceptable for many RCC products.

4.E.1. Digital Clock Synchronization Indicator

There are numerous ways that a RCC with the ability to display alphanumeric characters or symbols can provide a synchronization indicator. All are acceptable if they are clearly explained and understandable to the consumer. Figure 7 shows a watch displaying the date and time of the last synchronization. This is the preferred method since it passes the most information along to the consumer.

4.E.2. Analog Clock Synchronization Indicator

If an analog clock does not have a digital inset, a synchronization indicator should still be provided. For example, the indicator can be a light that is illuminated when synchronization has not occurred within the last 24 hours. Or advancing the second hand every two seconds instead of every second could be used to indicate that the clock has not synchronized in the last 24 hours. Other methods can be used, including the use of audio beeps or tones.

4.F. Adjustment of Display to Compensate for Delays Introduced During Synchronization

We recommend that manufacturers compensate for delays introduced during the synchronization of the display. For example, a RCC will require some processing time to decode the time code and to send the information to the display. The response time of the display will then introduce additional delays, which can be particularly large in the case of analog clocks that use a stepping motor. In some cases, the combined processing/response delays will exceed 0.1 s and become noticeable. The manufacturer should attempt to measure these delays and to advance the displayed time to compensate for the synchronization delays.

Manufacturers and consumers should also be aware of the path delay as the signal travels from WWVB to the RCC. However, this delay is relatively small; radio signals travel at the speed of light, and the path delay should never exceed 0.02 s within the WWVB coverage area. For this reason, and because path delay cannot be estimated without knowing the RCC's location, no path delay compensation is necessary.

5. RECOMMENDED PRACTICES FOR TIME ZONE SETTINGS

WWVB broadcasts UTC as opposed to local time. Therefore, each RCC must have a time zone switch or control that allows the local time zone to be selected in order for the clock to display local time.

5.A. Time Zone Selection

As a minimum requirement, all WWVB products should be capable of setting to the seven time zones listed in Table 3, so they can adequately service all potential consumers in the United States. However, since some WWVB products will be used through manual synchronization outside the coverage area (Section 4.D) or are capable of receiving other time signal stations (Section 3.D), we recommend that the time zone settings include each of the 38 time zones listed in Table 4. Both tables include the name, letter designation, and abbreviation for each time zone (when information is available), as well as the offset in hours from UTC.

Please note that if a RCC has the ability to select time zones offset by ±12 hours from UTC in half-hour increments, then nearly all of the time zones in the world will be represented. Figure 8 provides a world time zone map; Figure 9 provides time zone information for the United States. Manufacturers should consider including similar maps in their product documentation (Section 9).

Table 3: **Necessary Time Zone Options for WWVB RCC Products**

UTC Offset (hours)	Letter	Abbreviation	Name	Non-United States Areas Include
−10 :00	W	HST or HAST	Hawaii–Aleutian Standard Time	Central French Polynesia, Tokelau, Cook Islands, Tahiti, Johnston Atoll
−9 :00	V	AKST	Alaska Standard Time	Gambier Islands (French Polynesia)
−8 :00	U	PST	Pacific Standard Time	Western Canada, North Baja Peninsula
−7 :00	T	MST	Mountain Standard Time	West Central Canada, South Baja Peninsula, Central and Western Mexico
−6 :00	S	CST	Central Standard Time	Mexico, Easter Island, Galapagos Islands, Central Canada, Central America
−5 :00	R	EST	Eastern Standard Time	Western South America, Cuba, Bahamas, Haiti, Jamaica, East Central Canada, Panama
−4 :00	Q	AST	Atlantic Standard Time	Central South America, Dominican Republic, Eastern Canada, Puerto Rico, West Greenland, Bermuda

Table 4: **Recommended Time Zone Options for WWVB RCC Products Sold Internationally**

UTC Offset (hours)	Letter	Abbreviation	United States Name	Other Areas
−12 :00	Y			International Date Line West
−11 :00	X	SST	Samoa Standard Time	Midway Islands
−10 :00	W	HST or HAST	Hawaii-Aleutian Standard Time	Central French Polynesia, Tokelau, Cook Islands, Tahiti, Johnston Atoll
−9:30	—			Marquesas Islands (French Polynesia)
−9 :00	V	AKST	Alaska Standard Time	Gambier Islands (French Polynesia)
−8 :00	U	PST	Pacific Standard Time	Western Canada, North Baja Peninsula
−7 :00	T	MST	Mountain Standard Time	West Central Canada, South Baja Peninsula, Central and Western Mexico
−6 :00	S	CST	Central Standard Time	Mexico, Easter Island, Galapagos Islands, Central Canada, Central America
−5 :00	R	EST	Eastern Standard Time	Western South America, Cuba, Bahamas, Haiti, Jamaica, East Central Canada, Panama
−4 :00	Q	AST	Atlantic Standard Time	Central South America, Dominican Republic, Eastern Canada, Puerto Rico, West Greenland, Bermuda

Table 4: **Recommended Time Zone Options for WWVB RCC Products Sold Internationally** *(continued)*

UTC Offset (hours)	Letter	Abbreviation	United States Name	Other Areas
−3:30	—			Newfoundland Canada
−3:00	P			Eastern South America, Central Greenland
−2:00	O			Pernambuco (Brazil)
−1:00	N			Azores, East Greenland (Svalbard and Jan Mayen), Cape Verde
0	Z			Western Europe, Iceland, West Africa, Canary Islands, Coordinated Universal Time, Greenwich Mean Time
+1:00	A			Central Europe (including France and Spain), West Central Africa, Norway, Sweden, Denmark
+2:00	B			Eastern Europe, Russia Zone 1, East Central Africa, Turkey, Syria, Jordan, Greece, Cyprus, Israel, Lebanon, Finland
+3:00	C			Russia Zone 2, East Africa, Iraq, Saudi Arabia, Madagascar, Somalia, Sudan, Kuwait, Uganda, Yemen
+3:30	—			Iran

Table 4: **Recommended Time Zone Options for WWVB RCC Products Sold Internationally** *(continued)*

UTC Offset (hours)	Letter	Abbreviation	United States Name	Other Areas
+4 :00	D			Russia Zone 3, Georgia, Oman, Reunion, Mauritius, Seychelles, Azerbaijan, Armenia, United Arab Emirates
+4:30	—			Afghanistan
+5 :00	E			Russia Zone 4, British Indian Ocean Territory (Chagos), Kerguelen Island, Maldives Islands, Turkmenistan, Tajikistan, Pakistan, Uzbekistan, Western Kazakhstan
+5:30	—			India, Sri Lanka
+5:45	—			Nepal
+6 :00	F			Russia Zone 5, Eastern Kazakhstan, Bangladesh, Bhutan, Kyrgyzstan
+6:30	—			Cocos Islands, Burma
+7 :00	G			Russia Zone 6, Western Indonesia, Southeast Asia
+8 :00	H			Russia Zone 7, Western Australia, China, Hong Kong, Malaysia, Philippines, Central Indonesia, Singapore, Mongolia, Taiwan

Table 4: **Recommended Time Zone Options for WWVB RCC Products Sold Internationally** *(continued)*

UTC Offset (hours)	Letter	Abbreviation	United States Name	Other Areas
+9:00	I			Russia Zone 8, Japan, Korea, Palau, Eastern Indonesia
+9:30	—			Central Australia
+10:00	K	ChST	Chamorro Standard Time	Russia Zone 9, Eastern Australia, Chamorro (Guam and N. Mariana Islands), Micronesia, Papua New Guinea
+10:30	—			Lord Howe Island
+11:00	L			Russia Zone 10, Vanuatu, Solomon Islands, New Caledonia, East Micronesia Islands
+11:30	—			Norfolk Island
+12:00	M			Russia Zone 11, New Zealand, Fiji, Tuvalu, Marshall Islands, Nauru, Wake Island, Wallis and Futuna, Gilbert Islands (Kiribati), International Date Line East
+12:45	—			Chatham Islands
+13:00	—			Tonga, Phoenix Islands (Kiribati)
+14:00	—			Line Islands (Kiribati)

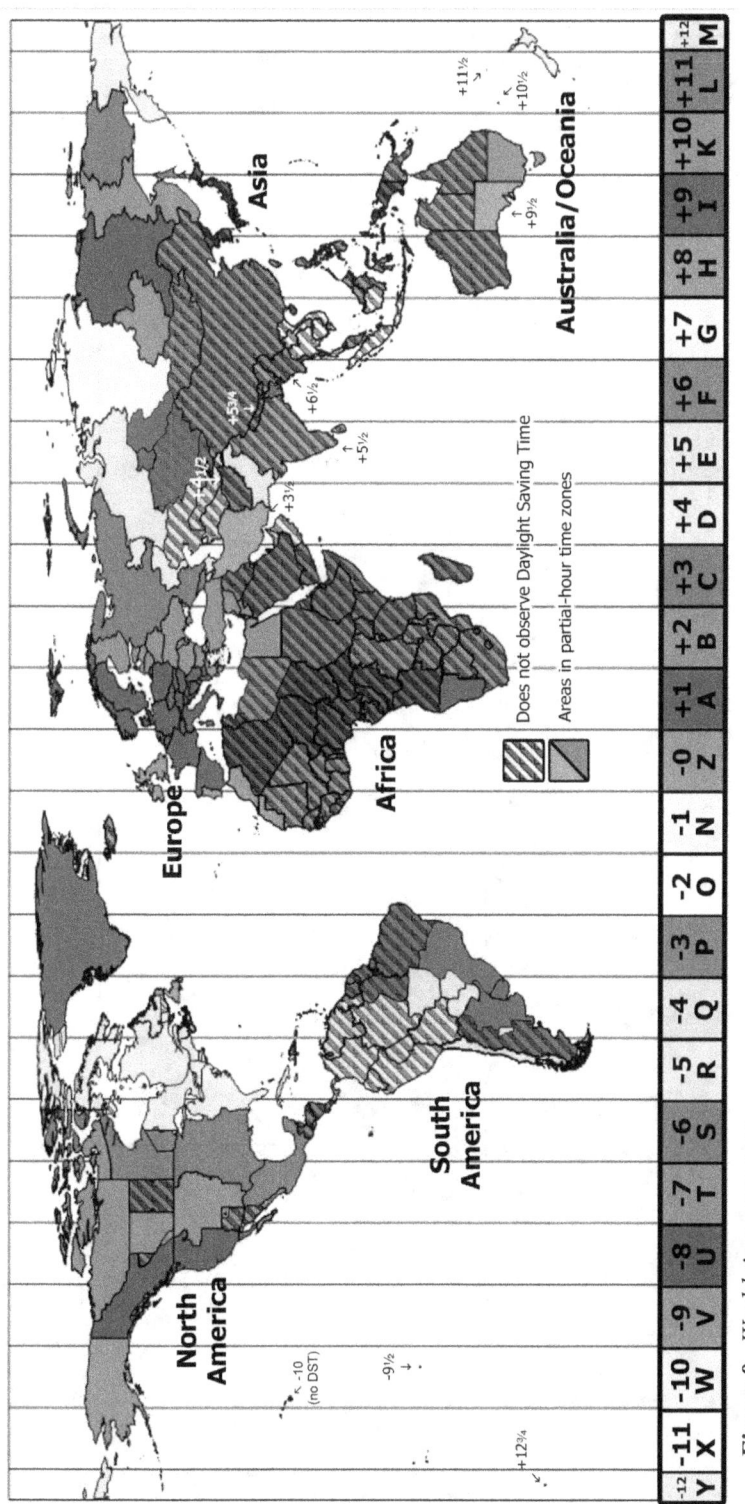

Figure 8. World time zone map.

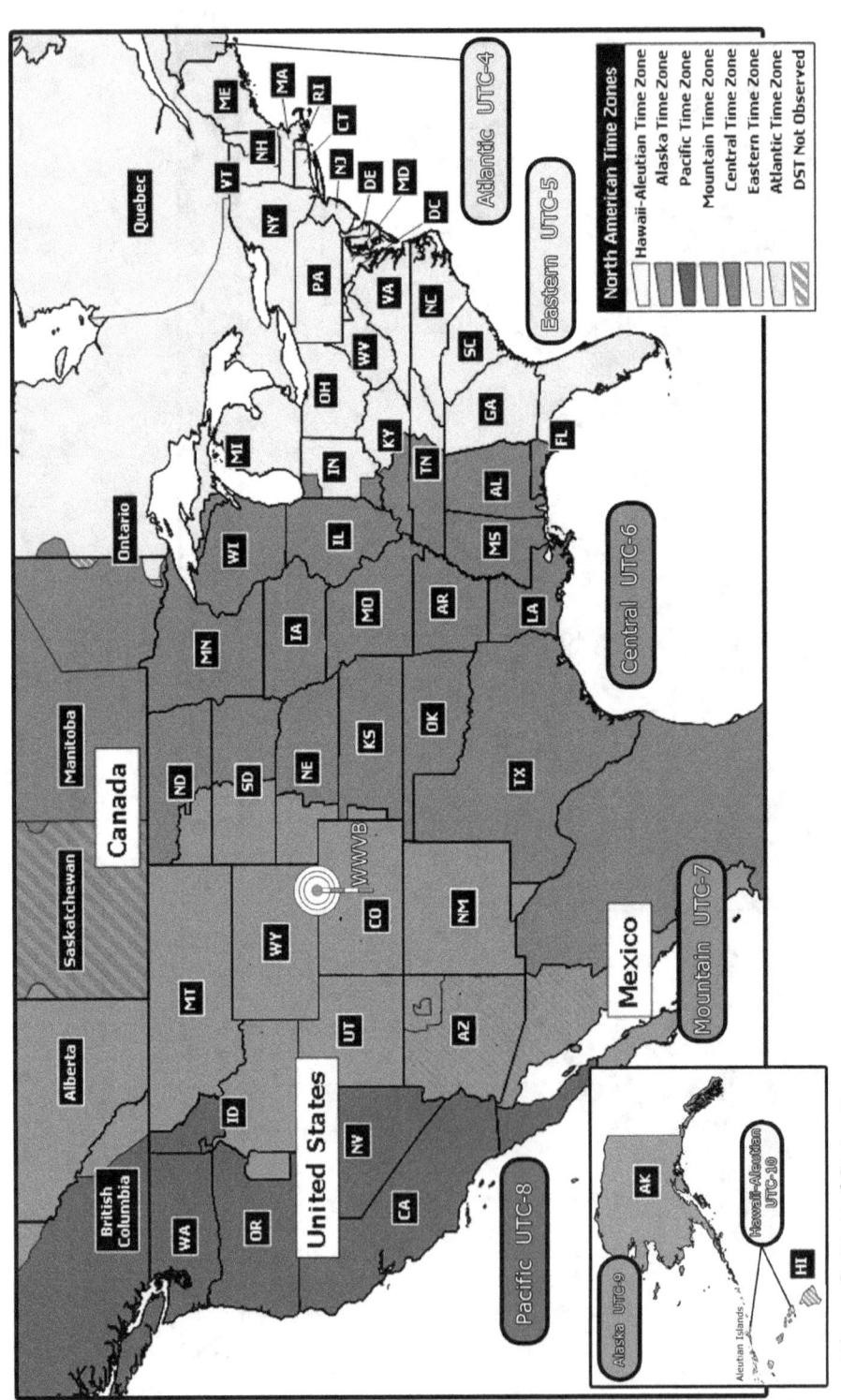

Figure 9. *United States time zone map.*

6. RECOMMENDED PRACTICES FOR DAYLIGHT SAVING TIME (DST)

The WWVB time code (Figure 1) includes information that tells the RCC whether standard time (ST) or daylight saving time (DST) is currently in effect, and also whether the current day is a transition day from ST to DST, or from DST to ST. All RCC products should decode this information so that consumers do not have to reset their clocks on the day of a time change. Figure 9 shows the areas within the United States that currently observe DST.

6.A. Handling of Transition Days

RCCs in areas that observe DST should advance one hour at 2 a.m. local time on the second Sunday of March each year, and move back one hour at 2 a.m. local time on the first Sunday of November of each year.[3] RCCs must properly interpret the information in the time code (Figure 1) and apply the time zone settings (Section 5), so the transition takes place at exactly 2 a.m. local time. Special attention to the DST code is required when implementing this feature as some manufacturers have misinterpreted the code. It might be necessary for analog clocks to move forward 11 hours (rather than back one hour) on the transition from DST to ST, if their clock mechanisms do not allow the hands to be moved backwards.

Manufacturers might elect to design products that implement the DST rule[3] at the assigned time, even if the RCC has been unable to recently read the DST information in the WWVB time code. This will allow the RCC to handle the DST transition even if the signal has not been recently received. However, if the DST rules change (they were changed in 1986 and 2007 and will probably change again), clocks programmed to follow them will fail, whereas the WWVB time code will always comply with the current rules. Therefore, we recommend that the DST information in the time code be used whenever possible and that the programmed rule only be used for backup or verification.

Another less important issue is that WWVB might be received under optimal conditions in areas (including South American countries such as Chile) that do not follow the same DST rules as the United States. This means RCCs located in these regions would be wrong at certain times of the year, regardless of whether the WWVB time code or the programmed DST rule for the United States was used to select the DST transition days. If a manufacturer sells products to consumers in these areas, we recommend that they allow the transition days and times to and from DST to be selected by the consumer through some type of user interface. Inclusion of this feature also guards against product obsolescence if the DST rules in the United States are changed again.

6.B. Disabling/Enabling DST Switch

Certain regions of the United States do not observe DST (Figure 9). Therefore, all RCC products must provide the consumer with the option to disable DST so that clocks located in those regions remain on ST throughout the year. Care should be taken to inform the consumer that this is normally a one-time setting; DST should be disabled only if their area does not observe DST, and not simply if ST is currently in effect.

6.C. DST Indicator

We recommend that the RCC has a display or switch setting that informs the consumer whether DST or ST is currently in effect.

7. RECOMMENDED PRACTICES FOR LEAP SECONDS, LEAP YEARS, AND THE TWO-DIGIT YEAR CODE

This section discusses the handling of miscellaneous time code settings, in particular the handling of leap seconds, leap years, and the two-digit year code.

7.A. Handling of Leap Seconds

When necessary (typically less than once per year), leap seconds are inserted into the UTC time scale on June 30th and/or December 31st. This keeps UTC within ±0.9 s of an astronomical time scale called UT1. The WWVB time code (Figure 1) includes a leap second bit (transmitted at second 56) that indicates whether a leap second will occur at the end of the current month. This allows the RCC to automatically insert the leap second. When a leap second does occur, the final minute of the day has 61 seconds. The UTC sequence looks like this:

23 hours, 59 minutes, 59 seconds

23 hours, 59 minutes, 60 seconds

0 hours, 0 minutes, 0 seconds

In order to properly display the occurrence of a leap second, digital RCCs must be capable of displaying a value of 60 in the seconds' field so that leap seconds can be indicated, for example 11:59:60 p.m. Analog RCCs cannot display a minute containing 60 seconds; therefore the second hand must remain in the same position (pointed straight up) for two consecutive seconds to indicate that a leap second has elapsed.

7.B. Handling of Leap Years

The WWVB time code (Figure 1) contains a leap year indicator (transmitted at second 55) that indicates whether the current year is a leap year. All RCCs that display date information must decode the leap year indicator so that dates after February 28th are correctly displayed. For example, if the RCC assumes that a leap year is a non-leap year, it will display a date of March 1st on February 29th.

The leap year rule is simple, and manufacturers can elect to design products that automatically know whether a year is a leap year by looking at the year code. However, since the WWVB time code uses two-digit year codes (Section 7.C), these must first be converted to four digits before testing to see whether the year is a leap year, due to the problem with centurial years (years that end with 00). Year numbers that are evenly divisible by 4 are leap years, with the exception of centurial years, which must be evenly divisible by 400.[4] Thus, 2000 was a leap year, whereas 2100, 2200, and 2300 will not be leap years. It is highly unlikely that the leap year rule will change, but in the event that it did, we recommend that the leap year information in the time code be used when possible. If the leap year rule is programmed into the product, we recommend that it be used only to backup or verify the time code information.

7.C. Handling of Two-Digit Year Code

The WWVB time code (Figure 1) includes only two digits of year information, so year information is ambiguous to the century. For example, the year 2004 is represented by "04." This is generally not a problem for most RCCs since a twenty-first century consumer viewing the clock will intuitively know that "04" means 2004. However, if the RCC is designed to interface with other devices, such as computer systems, the manufacturer should convert the year information to four digits. Simply adding 2000 to the two-digit year code will keep the four-digit year correct until the year 2100. While this seems like a good solution in the early part of the 21st century, other solutions should be sought if the RCC is expected to output the correct year information indefinitely.

8. RECOMMENDED PRACTICES FOR HARDWARE SPECIFICATIONS

This section discusses hardware specifications for the receiver, antenna, and local oscillator.

8.A. Receiver Specifications

Complete receiver specifications are beyond the scope of this document, but the minimum goal of the manufacturer should be to include a receiver and antenna sufficiently sensitive to work anywhere within the CONUS during the nighttime hours. We recommend that RCC products should be sensitive enough to successfully synchronize to signals from WWVB with a field strength of 50 µV/m, if the signal to noise ratio exceeds 20 dB. The RF bandwidth of the receiver should be narrow, typically ±10 Hz or less.

We recommend that digital signal processing (DSP) firmware be included in the receiver design to improve the RCC's ability to read the time code. The redundancy of the time code information (Figure 1) can be used to considerably improve reception. During a given hour, only the minute information in the time code changes from frame-to-frame (except during the rare hours when leap second, DST, or UT1 information happens to be inserted or deleted). Therefore, the time code normally changes from frame-to-frame in an entirely predictable fashion, which makes bit and frame averaging possible and desirable. When properly implemented, the use of bit and frame averaging can be more effective than increasing the sensitivity of the receiver, or increasing the field strength presented to the RCC by many decibels.

8.B. Antenna Considerations

Although external antennas obviously can provide better reception, we recommend that antennas be embedded inside the casing of the RCC to make the form factor more attractive and to prevent the antenna and/or its connecting wires from being damaged when the device is moved. Wristwatch antennas should not be contained in the band, so that RCC watch bands can be replaced in the same manner as the bands of ordinary watches when they are damaged or worn out. While external antennas are not recommended for stand-alone clocks and watches, they might be desirable for RCCs embedded in other devices, such as appliances.

8.C. Local Oscillator Specifications

As recommended in Section 4.B, all RCCs will attempt to synchronize at least once per day. Therefore, in order to meet the Section 3 requirement of keeping time between synchronizations to within ±0.5 s of UTC(NIST), the local quartz crystal oscillator must keep time to within about 0.48 s (we allow an 0.02 s error for signal propagation and internal clock synchronization delays) during a typical synchronization interval of 24 hours (86400 s). Therefore, the maximum allowable frequency offset of the quartz crystal oscillator, given a time change Δt in a period T, can be calculated as:

$$\frac{\Delta t}{T} = \frac{0.48}{86400} = 5.55 \times 10^{-6}$$

If the RCC uses a crystal oscillator with a nominal frequency of 32768 Hz, the maximum allowable frequency offset from nominal is about 0.18 Hz. If the synchronization period is shortened, for example if the RCC is able to synchronize every 12 h instead of every 24 h, these requirements are relaxed. Reducing the synchronization interval by a factor of 2 (from 24 h to 12 h) would double the maximum allowable frequency offset to 0.36 Hz. Care should be taken by the manufacturer to choose a quartz crystal oscillator that is accurate and stable enough to stay within its allowable tolerance over its normal operating temperature range, without requiring adjustment during the expected lifetime of the RCC.

Manufacturers may also choose to employ schemes that digitally compensate for the frequency offset of the quartz crystal, allowing the ±0.5 s/day specification to be met with a less stable oscillator.

8.D. Battery Powered RCCs

If a RCC product is battery powered, we recommend that a "low battery" indicator be included on the display, so the consumers are aware that the battery or batteries need to be changed. When the voltage of a battery-powered device drops below a certain threshold determined by the manufacturer, the clock should stop completely rather than attempt to keep time with an insufficient power supply.

9. RECOMMENDED PRACTICES FOR PRODUCT DOCUMENTATION

Although WWVB RCCs are less complicated than many other consumer electronic products, they do require more documentation than conventional clocks. An instruction sheet or manual that describes how to use all of the product's features should be included with every product sold. We also recommend that all buttons and controls on RCC products be clearly labeled; for example, identify the button(s) or control(s) used to change time zones. The documentation (preferably on the packaging itself) should indicate which time zones are supported by the product, and the approximate coverage area, so that consumers do not mistakenly buy products that won't work in their area. When space allows, such as on the back of a wall clock, it is helpful to engrave or stamp a condensed instruction sheet on the product itself.

Manufacturers are also encouraged to include text in their product documentation that describes how to troubleshoot RCC reception problems. Some examples are provided in Section 11 of this guide.

In numerous cases, consumers have complained that the loss of an instruction manual has made their product unusable. Therefore, we recommend that all instruction manuals for current and past models be made available to consumers on-line, where they can be downloaded free of charge if necessary.

9.A. Mention of NIST

We recommend that instruction manuals for WWVB RCC products mention that National Institute of Standards and Technology (NIST) radio station WWVB, located in Fort Collins, Colorado, is the source of the time signal received by the clock. It also should be noted that NIST is an agency of the United States government that provides official time to the United States. If manufacturers wish to provide a point of contact for obtaining more information about WWVB, they should reference the NIST Time and Frequency Division web site at **http://tf.nist.gov**. However, manufacturers should not direct consumers to NIST for technical support since NIST is unable to provide it.

9.B. Use of "Atomic Clock" Nomenclature

Many WWVB RCC products are labeled (on the product itself or in the documentation) as "atomic clocks." This is probably seen by manufacturers as a useful marketing tool intended to capture the imagination of potential customers, and some might argue that it is appropriate since atomic clocks are located at the WWVB radio transmitter site. However, we contend that use of the term "atomic clock" is technically incorrect and misleading to consumers, and its usage should be avoided. Unless there is actually an atomic oscillator inside the RCC (such as a cesium or rubidium oscillator), we recommend that the term "radio controlled clock" be used to correctly describe the product. Labeling products or documentation with the term "atomic timekeeping" is also considered acceptable.

10. COMPLIANCE CHECKLIST

The checklist below is included to assist both manufacturers and consumers of RCC products. By answering the questions in the table, it can quickly be determined whether or not a given product complies with the recommendations provided in this handbook.

Section	Question	N = Necessary O = Optional	Product Complies with Recommendation (Preferred Answer Is Always Yes)	
			Yes	No
3	Is time always displayed to within ±0.5 s during the entire interval between clock synchronizations?	N		
3	Is time always displayed to within ±0.2 s during the entire interval between clock synchronizations?	O		
3	Does the clock display seconds?	N		
3.A/3.B	Does the clock have a label or icon indicating that it is radio controlled?	N		
3.A/3.B	Does the clock display the date?	O		
3.B	If the clock is digital, does it include a.m. and p.m. indicators?	O		
3.B	Was the clock designed without an icon or picture of a satellite dish on its display?	O		
3.C	Are the clock controls clearly labeled and situated?	N		
3.D	If the clock is capable of receiving more than one time signal station, does it indicate which station it is receiving?	O		
3.E	Does the clock have a way to indicate signal quality?	O		

Section	Question	N = Necessary O = Optional	Product Complies with Recommendation (Preferred Answer Is Always Yes)	
			Yes	No
3.E	Is the signal quality meter not visible or disabled when the clock is not attempting to synchronize?	O		
3.F	Does the case have a marker indicating the orientation of the antenna?	O		
4.A	Does the clock continuously try to synchronize when it is first turned on?	N		
4.A	Does the clock refrain from displaying the time prior to its first synchronization?	N		
4.A	Does the clock decode and compare at least two consecutive time codes before determining that the initial synchronization attempt is successful?	N		
4.A	Can the clock synchronize without having the display or hands preset by the consumer?	N		
4.B	Does the clock attempt to synchronize by radio at least once every 24 hours?	N		
4.B	Does a clock that has been previously synchronized continue to keep time during a synchronization attempt?	N		
4.B	Does the clock attempt to synchronize during the nighttime hours when the signal from WWVB is the strongest?	N		

Section	Question	N = Necessary O = Optional	Product Complies with Recommendation (Preferred Answer Is Always Yes)	
			Yes	No
4.B	Does the clock attempt more than one synchronization every 24 hours?	O		
4.B	Does the clock allow at least five minutes for a synchronization attempt?	N		
4.C	Does the clock include a button or control that allows the consumer to attempt to synchronize at any time?	N		
4.D	Can the clock be set manually, without radio synchronization?	N		
4.E	Does the clock include a synchronization indicator?	N		
4.F	Does the clock compensate for delays introduced during the synchronization of its display?	N		
5.A	Does the clock allow each of the time zones listed in Table 3 to be selected?	N		
5.A	Does the clock allow each of the time zones listed in Table 4 to be selected?	O		
6	Does the clock automatically adjust on the transition days from ST to DST, and from DST to ST?	N		
6.A	Does the clock change from standard time to DST at 2 a.m. local time and vice versa?	N		

Section	Question	N = Necessary O = Optional	Product Complies with Recommendation (Preferred Answer Is Always Yes)	
			Yes	No
6.B	Does the clock have a way to disable DST for areas that do not observe it?	N		
6.C	Does the clock include a DST indicator?	O		
7.A	Does the clock properly handle leap seconds?	O		
7.B	If the clock displays date information, does it properly handle leap years?	N		
7.C	If the clock displays four-digit year information, does it properly handle the two-digit year code?	O		
8.A	Does the clock's receiver meet specifications?	N		
8.B	Is the antenna concealed inside the clock unit?	O		
8.C	Does the local oscillator meet specifications?	N		
8.D	If the clock is powered by batteries, is a "low battery" indicator included?	O		
9	In the case of wall clocks, are condensed instructions included on the product itself?	O		
9	Does the manufacturer make instructions manuals for the clock available on-line?	O		

Section	Question	N = Necessary O = Optional	Product Complies with Recommendation (Preferred Answer Is Always Yes)	
			Yes	No
9.A	Does the product documentation mention NIST and WWVB?	O		
9.B	Does the product documentation and/or labeling use the term "radio controlled clock" or "atomic timekeeping," instead of the incorrect "atomic clock?"	O		

11. RECOMMENDED PRACTICES FOR CONSUMERS OF WWVB RCCS

The first and foremost reason that consumers purchase RCCs is accuracy. A RCC has a tremendous advantage over a conventional clock: when working properly, it is always right! Consumers never need to adjust a RCC, not even during the transition between daylight saving time and standard time. However, this section covers a few items that consumers need to know about to ensure that their RCC is working properly and providing the correct time.

11.A. How a WWVB RCC Works

WWVB RCCs are conventional quartz clocks with a miniature radio receiver inside that is permanently tuned to receive the 60 kHz signal from NIST radio station WWVB (Section 2). This station is located near Fort Collins, Colorado, about 100 km north of Denver. It broadcasts a time signal continuously, 24 hours per day, 7 days per week, with a complete time message (called a time code) sent every minute. However, RCC products only attempt to read this time code periodically, often only once every 24 hours, typically during the night when the signal is strongest (Section 4.B).

The 60 kHz signal is located in a part of the radio spectrum called LF, which stands for low frequency. This is an appropriate name, because the FM radio and TV broadcasts that we are accustomed to listening to use frequencies thousands of times higher. The lowest frequency received by any other consumer radio is probably 530 kHz, the bottom of the AM broadcast band, and even that frequency is nearly nine times higher than the WWVB frequency.

The 60 kHz signal does not provide enough bandwidth to carry audio information. Instead, all that is sent is a time code. The time code is simply a message containing time and date information. This information is sent in the form of binary digits, or bits, which have two possible values (0 or 1). Frame markers are also sent as part of the message, so the RCC can align the time code and read the bits in their proper order. The time code bits are generated by raising and lowering the power of the WWVB signals. They are sent at a very slow rate of 1 bit per second, and it takes a full minute to send a complete time code or a message that tells the clock the current date and time. When a RCC is first turned on, it will probably miss the first time code, so it usually takes at least two minutes to synchronize, depending upon the signal quality and the receiver design.

11.B. Time Zone Settings

Since the WWVB signal originates from a single location in Colorado, it does not contain any time zone information. Therefore, WWVB RCCs can not determine which time zone they are in unless this information is supplied by the consumer. Well-designed products (Section 5) allow the selection of all time zones where the clock could possibly be used.

The time broadcast by WWVB is Coordinated Universal Time (UTC), or the time kept at the Prime Meridian that passes through Greenwich, England. Clocks all over the world are synchronized to the same second as UTC in all cases and the same minute as UTC in nearly all cases.* However, the local hour is different than the UTC hour, based on the number of time zones between the local time zone and the Prime Meridian. While a few consumers want their clocks to display UTC (ham radio operators, for example), most prefer to display local time. WWVB RCCs apply a time zone correction of the UTC hour to convert UTC to local time. The size of this correction is shown in Table 5 for the four major time zones in CONUS.

When consumers move a WWVB RCC to another time zone, they need to change the time zone setting accordingly. Consumers that travel with RCCs should familiarize themselves with the procedure for changing time zones, so they can adjust their clocks whenever necessary.

---

* A few time zones (Table 4) differ from UTC by a non-integer number of hours (3.5 hours, for example). Clocks synchronized to local time in these regions will display a different minute than a clock synchronized to UTC, but the second will be the same.

Table 5: **Difference between UTC and Local Time for the Four Major Time Zones in the CONUS**

Time Zone	Difference from UTC During Standard Time	Difference from UTC During Daylight Time
Pacific	−8 hours	−7 hours
Mountain	−7 hours	−6 hours
Central	−6 hours	−5 hours
Eastern	−5 hours	−4 hours

11.C. Coverage Area of the WWVB Signal

During the nighttime hours, the WWVB signal is strong enough to synchronize clocks in the 48 states of the CONUS, in parts of Alaska and Hawaii, in all of Mexico, in most of the populated areas of Canada, and in some regions of Central and South America. (For coverage maps and signal strength information recorded at various sites, see **http://tf.nist.gov**)

The size of the coverage area is estimated using a field strength figure of 100 μV/m, which in theory is more than enough signal for a well-designed RCC to synchronize (Section 8.A). However, in practice, simply having a large signal doesn't mean that a RCC will be able to work. What really matters is the signal-to-noise ratio, or the size of the signal compared to the size of the electrical noise near the same frequency. Raising the noise level is just as harmful as reducing the signal level. For example, if the RCC clock is near a source of interference, the noise level increases, and the clock might not be able to synchronize even if the local field strength of the time signal is high. Potential sources of interference are discussed in Section 11.D.1.

11.D. General Troubleshooting Tips for WWVB RCCs

WWVB RCC products have different specifications, and use different controls and user interfaces, so technical support must be provided by the manufacturer and not by NIST. We recommend that consumers save the instruction sheets that come with their clocks, so they can refer to them in the future if necessary. Having said that, this section offers a few general tips for consumers whose RCCs aren't displaying the correct time.

Nearly all problems reported by consumers with WWVB RCCs are related to the clock itself and not to the WWVB broadcast. Consumers should be aware that RCC problems caused by the WWVB broadcast are extremely rare. WWVB has a number of safeguards in place to help ensure that the correct time is always being broadcast, and time is kept at the station to within 100 nanoseconds of UTC.[2] The station does occasionally have signal outages, and all outages over five minutes in length are listed at **http://tf.nist.gov**. However, most outages are maintenance related and occur in the daytime hours when RCCs are not attempting to synchronize, so they have no effect on consumer products. Unplanned outages during the nighttime hours are responded to as quickly as possible and rarely last for more than 1 or 2 hours. While it is possible that one of these outages can cause a RCC to miss one daily synchronization period, it is highly unlikely that this will happen two days in a row. Field strength varies due to the time of year and the current weather conditions, but it should be sufficient for RCCs in the CONUS to synchronize during each night of the year.

Consumers who suspect that their WWVB RCC is not displaying the correct time can check it by comparing it to other NIST time services, including the NIST web clock **http://time.gov** or the audio time signals from NIST radio stations WWV and WWVH.[1] The audio time signals can be heard using a shortwave radio or by telephone (dial 303–499–7111). WWVB RCCs should be within ±0.5 s of either source. Please note that the NIST web clock allows time zone selection and displays local time, but consumers must convert from UTC to get local time from WWV or WWVH (Table 5).

11.D.1. General Troubleshooting Tips for RCCs That Won't Synchronize at All

If their RCC won't synchronize, we recommend that consumers try the following:

- If the RCC uses batteries, check them and replace if necessary. Low batteries can cause a variety of RCC problems. If the RCC used to work, but doesn't work now, try changing the batteries before deciding it has failed.

- If you have a desktop RCC, try rotating it 90°. If you have a wall clock, try mounting it on a wall perpendicular to the one it is currently on (*e.g.*, if it is on a north–south wall try an east–west wall). The antennas are directional, and reception can be improved by turning the antenna. If your RCC has a signal quality indicator and antenna orientation markers (Sections 3.E and 3.F), use them to help determine the proper antenna orientation.

- Place the RCC along a wall or near a window that faces Fort Collins, Colorado.

- If you are staying in a hotel and traveling with a WWVB alarm clock or wristwatch, it will probably work best if you leave it near the window overnight.

- If you have a WWVB wristwatch, remove it from your wrist at night so that it is motionless during the synchronization period (Section 4.B).

- Locate the clock away from the potential sources of interference listed in Table 6.

Table 6: **Potential Sources of Interference for WWVB RCCs**

Source of interference	Reason for interference
Computer monitors and televisions	Some monitors have a scan frequency at or near the WWVB carrier frequency of 60 kHz. Place RCCs at least 1 to 2 meters away from computer monitors or televisions for best results.
Metal or ferroconcrete buildings	Buildings made out of metal (such as mobile homes) or buildings with metal roofs or steel siding might prevent the clock from working by blocking or weakening the incoming signal. Ferroconcrete buildings (where the concrete has metal added to provide extra support), can also interfere with the signals. Place the clock near a window to give it the best chance of synchronizing.
Refrigerators, air conditioners, household appliances, or other devices with electric motors	Electric motors can generate radio frequency interference (RFI) at the AC line frequency of 60 Hz, which is a subharmonic of the 60 kHz WWVB carrier. Place the RCC at least 1 or 2 meters away from these devices for best results.
Basements or underground locations	WWVB signals can reach underground locations better than higher frequency signals, but the signal quality will be lessened if the RCC is placed in a basement. Also, basement walls are often made of ferroconcrete material (see above). Try placing the clock above the ground if it doesn't work in a basement location.
Neon or fluorescent lights	Neon or fluorescent lights can sometimes emit RFI that interferes with RCCs. If consumers suspect that lights are producing RFI, they can perform a simple test with a battery-powered AM radio. The radio should be tuned to a dial location between stations so that only noise is heard.

Table 6: **Potential Sources of Interference for WWVB RCCs** *(continued)*

Source of interference	Reason for interference
Neon or fluorescent lights *(continued)*	If possible, turn off the lights to see if this noise level goes down. If it is not possible to turn off the light, walk towards the light with the radio to see if the noise increases. Try placing the RCC where the noise level is lowest to see if that helps.
Electrical storms	Electrical storms can generate RFI in the part of the spectrum used by WWVB. Lightning along the path between the consumer's receiver and Fort Collins, Colorado, can potentially cause an RCC to miss a scheduled synchronization, but the problem should cease when the storm is over.
Overhead electrical equipment related to electrical power generation	The 60 Hz RFI emitted by electrical wires or other power generating equipment can be a problem since it is a subharmonic of the 60 kHz WWVB carrier. Try moving the RCC as far from the RFI source as possible.
Broadcasting stations	RCCs located near a radio or television station can be overloaded by the local signal. This can be a particular problem if 60 kHz is a subharmonic of the local carrier.

If nothing else works, we recommend that consumers take the clock outdoors after dark and power it down (remove the batteries or unplug it), then power it up again to force it to look for the WWVB signal. If it works outdoors but not indoors, there is probably a local interference problem inside the building where the clock is located. If it doesn't work outdoors at night, it's probably best to return it and try a different model.

Keep in mind that not all WWVB RCCs are created equal. Consumers who think their clock is defective or that it is simply unable to work at their location should ask the manufacturer or dealer for a replacement.

11.D.2. General Troubleshooting Tips for RCCs Off by One Hour or More

Remember, minutes and seconds are the same in nearly all time zones, only hours are different. Therefore, if a clock is off by exactly one hour or by a multiple of exactly one hour, it probably has to do with a time zone setting. Consumers should make sure that they have properly selected their local time zone using the instructions that came with the RCC. If the clock does not have a setting for the consumer's local time zone (Section 5.A), we recommend that they return it to the dealer. Consumers outside the CONUS should make sure that a RCC product handles their time zone prior to purchasing it.

Consumers who live in areas that do not observe Daylight Saving Time (Figure 9) must make sure that DST is disabled on their RCC (Section 6.B). If the RCC lacks this feature, consumers might still be able to select another time zone to make the RCC display the correct time when DST is in effect.

11.D.3. General Troubleshooting Tips for RCCs Off by a Few Minutes or Seconds

Properly working and designed RCCs should display time accurate to within ±0.5 s or better (Section 3). However, a malfunctioning RCC can be off by a few seconds or even minutes, for the reasons listed below:

- **Reception Problem** — If the RCC isn't receiving the signals from WWVB at least once per day, the time will "drift" and gradually get further and further from the correct time. Remember, if WWVB isn't being received, the clock is no longer radio controlled, it's just a conventional quartz clock. Its accuracy will then depend on the quality of the quartz crystal (Section 8.C). Most quartz clocks can keep time to 1 s per day or better, but some could be off by several seconds per day and a time error always accumulates between synchronizations. Consumers are advised to purchase only RCCs with synchronization indicators (Section 4.E), so they will know whether or not the clock has recently synchronized. If the product does not have a synchronization indicator and the consumer cannot tell if the signal is being received, we recommend powering down the clock (by unplugging it or removing the batteries), then powering it up again to see if it can synchronize. If it doesn't, see Section 11.D.1 for tips on improving reception.

- **Alignment Problem** — Consumers might obtain analog RCCs whose hands aren't properly aligned. This could cause the clock to be off by 1 s or more even if it is receiving the signal properly. The clock might not have been properly aligned at the factory, or it might have been jostled during shipment, causing the hands to move. Some manufacturers explain how to align the hands on their instruction sheet. If the consumer is unable to do this, and if the small error bothers them, we recommend that they return the clock to the dealer for replacement.

- **Parallax Problem** — The parallax problem refers to the apparent shifting of an object when viewed at different angles and can prevent problems when viewing analog RCCs. When consumers check the accuracy of an analog RCC, they need to be sure they are looking straight at the clock face and not viewing it from an angle. Consumers who view the clock from an angle might think it is off by a few seconds even if it is not. This is similar to trying to read the speedometer from the passenger seat of a car and thinking the speed is faster or slower than it actually is.

11.D.4. General Troubleshooting Tips Concerning Daylight Saving Time (DST)

Problems related to DST are among the most common problems experienced by RCC consumers. If a RCC doesn't change during the transition from DST to standard time, or vice versa, it probably means that it has not received the signal recently, so it didn't know about the time change. See Sections 11.D.1 and 11.D.3 for tips on improving reception.

If reception appears to be fine and the RCC didn't change, consumers should make sure that DST is not disabled on their RCC (Section 6.B), if their area observes DST. Conversely, if the area does not observe DST and the RCC did change, consumers need to disable the DST setting.

12. REFERENCES

[1] Lombardi, M.A. "NIST Time and Frequency Services." *Natl. Inst. Stand. Technol. Spec. Publ. 432,* 71 p., January 2002.

[2] Nelson, G.K., M.A. Lombardi, and D.T. Okayama. "NIST Time and Frequency Radio Stations: WWV, WWVH, and WWVB." *Natl. Inst. Stand. Technol. Spec. Publ. 250–67,* 156 p., January 2005.

[3] 15 United States Code 6(IX)(260a).

[4] Seidelmann, P. Kenneth, ed. "Explanatory Supplement to the Astronomical Almanac." *University Science Books,* pp. 580–581, 1992.

August 2009

www.ingramcontent.com/pod-product-compliance
Lightning Source LLC
Chambersburg PA
CBHW081737170526
45167CB00009B/3853